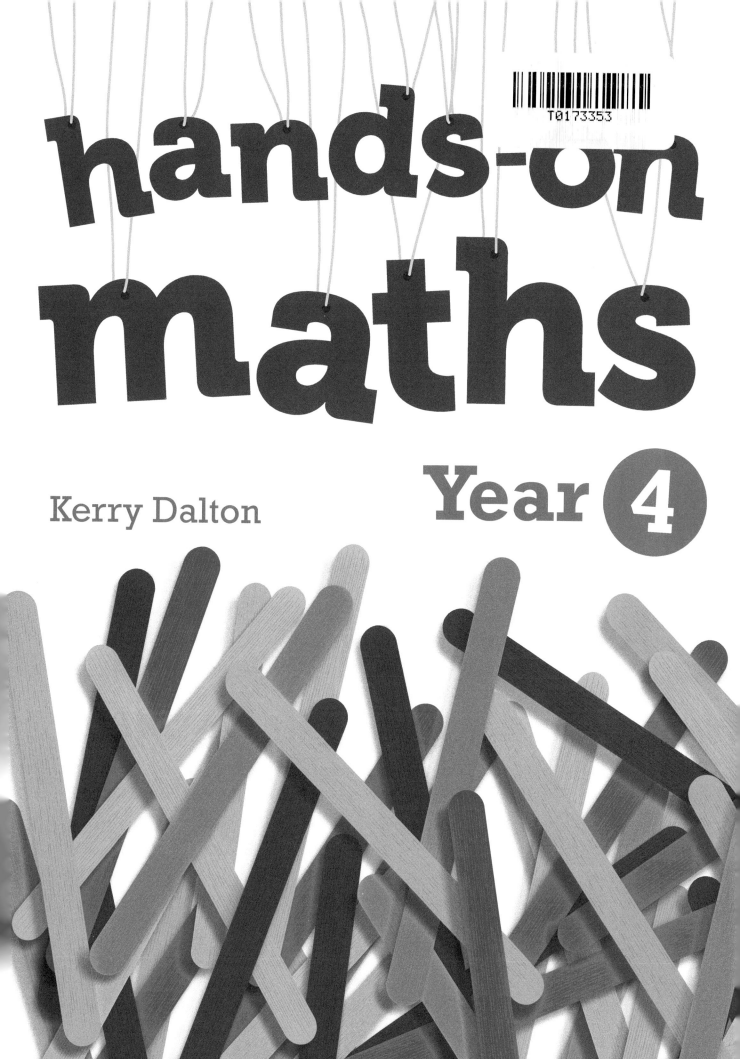

hands-on maths

Year **4**

Kerry Dalton

Published by Keen Kite Books
An imprint of HarperCollins*Publishers* Ltd
The News Building
1 London Bridge Street
London
SE1 9GF

ISBN 9780008266981

First published in 2017
10 9 8 7 6 5 4 3 2

Series Concept and Commissioning: Shelley Teasdale and Michelle I'Anson
Project Manager: Fiona Watson
Editor: Denise Moulton
Cover Design: Anthony Godber
Text Design and Layout: Contentra Technologies
Production: Natalia Rebow
A CIP record of this book is available from the British Library.

Contents

Year 4 aims and objectives

Hands-on Maths Year 4 encourages pupils to enjoy a range of mathematical concepts through a practical and hands-on approach. Using a range of everyday objects and common mathematical resources, pupils will explore and represent key mathematical concepts. These concepts are linked directly to the National Curriculum 2014 objectives for Year 4. Each objective will be investigated over the course of the week using a wide range of hands-on approaches such as Dienes, place-value counters, abacus models, playing cards, dice, place-value grids, practical problems and a mix of individual and paired work. The mathematical concepts are explored in a variety of contexts to give pupils a richer and deeper learning experience, which enables mastery to be attained.

Year 4 programme and overview of objectives

Topic	Week 1	Week 2	Week 3	Week 4	Week 5	Week 6
Counting	Count from 0 in multiples of 6	Count from 0 in multiples of 9	Count from 0 in multiples of 7	Count from 0 in multiples of 25	Count from 0 in multiples of 1000	Count backwards through 0 to include negative numbers
Place value	Recognise the place value of each digit in a four-digit number	Order and compare numbers beyond 1000	Find 1000 more or less than a given number	Solve number problems involving place value ideas with increasingly large positive numbers	Solve number problems involving place value ideas with increasingly large positive numbers	Solve practical problems involving place value ideas with negative numbers
Representing numbers	Identify, represent and estimate numbers using different representations	Read Roman numerals to 10	Read Roman numerals to 50	Read Roman numerals to 100	Round any number to the nearest 10 or 100	Round any number to the nearest 10, 100 or 1000

Year 4 aims and objectives

Topic	Week 1	Week 2	Week 3	Week 4	Week 5	Week 6
Addition and subtraction	Add mentally a four-digit number and ones	Subtract mentally a four-digit number and ones	Add mentally a four-digit number and tens / hundreds	Subtract mentally a four-digit number and tens / hundreds	Add mentally a four-digit number and thousands (finding 1000 more)	Subtract mentally a four-digit number and thousands (finding 1000 less)
Multiplication and division	Recall and use multiplication and division facts for the 6 multiplication table	Recall and use multiplication and division facts for the 9 multiplication table	Recall and use multiplication and division facts for the 11 multiplication table	Recall and use multiplication and division facts for the 12 multiplication table	Solve problems involving multiplication and division, with 0 and 1	Recognise and use factor pairs and commutativity in mental calculations
Fractions	Count up and down in hundredths; recognise that hundredths arise when dividing an object by 100 and dividing tenths by 10	Recognise and show, using diagrams, families of common equivalent fractions	Add and subtract fractions with the same denominator	Recognise and write decimal equivalents of any number of tenths or hundredths	Recognise and write decimal equivalents of any number of tenths and hundredths, and to $\frac{1}{4}$, $\frac{1}{2}$, $\frac{3}{4}$	Round decimals with one decimal place to the nearest whole number

Introduction

The *Hands-on maths* series of books aims to develop the use of readily available manipulatives such as toy cars, shells and counters to support understanding in maths. The series supports a concrete–pictorial–abstract approach to help develop pupils' mastery of key National Curriculum objectives.

Each title covers six topic areas from the National Curriculum (counting; representing numbers; place value; the four number operations: addition and subtraction and multiplication and division; and fractions). Each area is covered during a six-week unit, with an easy-to-implement 10-minute activity provided for each day of the week. Photos are included for each activity to support delivery.

Hands-on maths enables a deep interrogation of the curriculum objectives, using a broad range of approaches and resources. It is not intended that schools purchase additional or specialist equipment to deliver the sessions; in fact, it is hoped that pupils will very much help to prepare resources for the different units, using a range of natural, formal and typical maths resources found in most classrooms and schools. This will help pupils to find ways to independently gain a deep understanding and enjoyment of maths.

A typical 'hands-on' classroom will have a good range of resources, both formal and informal. These may include counters, playing cards, coins, Dienes, dominoes, small objects such as toy cars and animals, Cuisenaire rods, 100 squares and hoops.

There is no requirement to use *only* the resources seen in the photographs that accompany each activity. Cubes may look like those in the green bowl, or will be just as effective if they look like the ones in the blue bowl. They serve the same purpose in helping pupils understand what the cubes represent.

Resources

Hands-on maths uses a range of formal, informal and 'typical' resources found in most classrooms and schools. To complete the activities in this book, it is expected that teachers will have the following resources readily available:

- whiteboards and pens for individual pupils and pairs of pupils
- Dienes and Cuisenaire rods
- dice, coins and bead strings
- a range of cards, including playing cards, place-value arrow cards and digit cards

- collections of objects that pupils are interested in and want to count, such as toy cars, toy animals and shells
- bowls / containers to store sets of resources in, making it easy for pupils to handle and use the objects

- ten frames (these could be egg boxes, ice-cube trays, printed frames or something pupils have created themselves)
- number lines and 100 squares – lots of different types and styles: printed, home-made, interactive, digital or practical … whatever you prefer, and whatever is handy. (For 100 squares, there is, of course, the 1–100 or 0–99 choice to make; both work and it is best to choose whatever works for the class. Both offer a slight difference in place-value perspective, with 0–99 giving the 'zero as a place holder' emphasis, while the 1–100 version helps pupils to visualise the position of 100 in relation to the two-digit numbers.)

- counters and cubes – lots of them! Many of the activities require counters and cubes to be readily available. The cubes can be any size and any colour: what the cubes represent is the most important factor.

Maths is a truly unique, creative and exciting discipline that can provide pupils with the opportunity to delve deeply into core concepts. Maths is found all around us, every day, in many different forms. It complements the principles of science, technology and engineering.

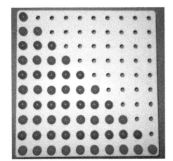

Hands-on maths provides ideas that can be adapted to suit the broad range of needs in our classrooms today. These ideas can be used as a starting point for assessment – before, during or after teaching a particular topic has taken place. The activities are intended to be flexible enough to be used with a whole class and can, of course, be differentiated to suit individual pupils in a class.

The activities can be adapted to link to other subject areas and interests. For example, a suggestion to use farm animals may link well to a science unit on classification or food chains; alternatively, the resource could be substituted with bugs if minibeasts is an area of interest for pupils. Teachers can be as flexible as they wish with the activities and resources – class teachers know their pupils best.

Spoken language is underpinned in maths by the unique mathematical vocabulary pupils need to be able to use fluently in order to demonstrate their reasoning skills and show mathematical proof. The correct, regular and secure use of mathematical language is key to pupils' understanding; it is the way in which they reason verbally, negotiate conceptual understanding and build secure foundations for a love of mathematics and all that it brings. Each unit in *Hands-on maths* identifies a range of vocabulary that is typical, but by no means limited to, that particular unit. The way the vocabulary is used and incorporated into activities is down to individual style and preference and, as with all of the resources in the book, will be very much dependent on the needs of each individual class. A blank template for creating vocabulary cards is included at the back of this book.

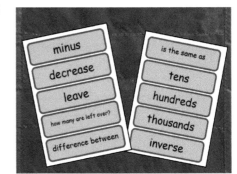

Week 1: Counting

Count from 0 in multiples of 6

Resources: 2cm squared paper, colouring pencils, scissors, envelopes, 100 squares, number lines, sticky notes, playing cards

> **Vocabulary:** counting, number, zero, one, two, three …, ten, twenty … one hundred, two hundred … one thousand, how many …?, count, count (up) to / on / back, count in … sixes, sevens, nines, twenty-fives, thousands, more, less, many, few, tally, odd, even, every other, skip count, how many times?, multiple of, sequence, continue, predict, pattern, pair, rule, relationship

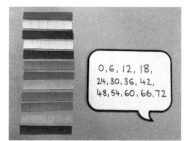

Monday

Give each pupil a sheet of 2 cm squared paper and some colouring pencils.

Ask if pupils can think of things that come in sixes (e.g. boxes of eggs). Each pupil colours the squares in strips of 6, then cuts out 12 strips and places them in a named envelope for use throughout the week. (This activity would work equally well with objects grouped into sixes.)

Ask pupils to lay out their strips. Then count together from 0–72, pointing to each strip of 6 as they count. Practise counting both forwards and backwards.

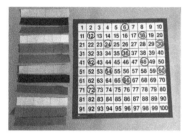

Tuesday

Give each pupil a 100 square and their set of strips.

Repeat Monday's activity. This time, when you say out loud a multiple of 6, pupils circle that number on their 100 square. Count both forwards and backwards.

Wednesday

Give each pupil or pair of pupils a number line from 0–100 and their set of strips.

Say out loud a question from the 6 multiplication table and ask pupils to find that multiple of 6 using the strips. They should then circle the total on their number line. Count both forwards and backwards.

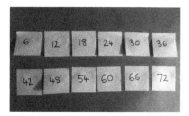

Thursday

Give each pupil or pair of pupils 12 sticky notes (or small squares of paper).

Ask pupils to write the multiples of 6 from 6–72 on the notes and then to place them in order from smallest to largest. Count forwards and backwards using the numbers as a resource. Keep the resource for Friday.

Friday

Pupils again lay out their sticky notes in order from smallest to largest. Show a playing card. Pupils count in sixes that number of times and hold up the note which is the multiple shown on the card.

Week 2: Counting

Count from 0 in multiples of 9

Resources: 2cm squared paper, colouring pencils, scissors, envelopes, 100 squares, number lines, sticky notes, playing cards

Vocabulary: counting, number, zero, one, two, three …, ten, twenty … one hundred, two hundred … one thousand, how many …?, count, count (up) to / on / back, count in … sixes, sevens, nines, twenty-fives, thousands, more, less, many, few, tally, odd, even, every other, skip count, how many times?, multiple of, sequence, continue, predict, pattern, pair, rule, relationship

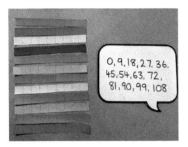

Monday

Give each pupil a sheet of 2cm squared paper and some colouring pencils.

Each pupil colours the squares in strips of 9, then cuts out 12 strips and places them in a named envelope for use throughout the week. (This activity would work equally well with objects grouped into nines.)

Ask pupils to lay out their strips. Then count together from 0–108, pointing to each strip of 9 as they count. Practise counting both forwards and backwards.

Tuesday

Give each pupil a 100 square and their set of strips.

Repeat Monday's activity. This time, when you say out loud a multiple of 9, pupils circle that number on their 100 square. Highlight that 11 × 9 is the last multiple that falls in the range 0–100. Count both forwards and backwards.

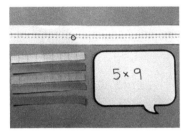

Wednesday

Give each pupil or pair of pupils a number line from 0–100 and their set of strips.

Say out loud a question from the 9 multiplication table and ask pupils to find that multiple of 9 using the strips. They should then circle the total on their number line. Count both forwards and backwards.

Thursday

Give each pupil or pair of pupils 12 sticky notes (or small squares of paper).

Ask pupils to write the multiples of 9 from 9–108 on the notes and then to place them in order from smallest to largest. Count forwards and backwards using the numbers as a resource. Keep the resource for Friday.

Friday

Pupils again lay out their sticky notes in order from smallest to largest. Show a playing card. Pupils count in nines that number of times and hold up the note which is the multiple shown on the card.

Week 3: Counting

Count from 0 in multiples of 7

Resources: 2cm squared paper, colouring pencils, scissors, envelopes, 100 squares, number lines, sticky notes, playing cards

Vocabulary: counting, number, zero, one, two, three …, ten, twenty … one hundred, two hundred … one thousand, how many …?, count, count (up) to / on / back, count in … sixes, sevens, nines, twenty-fives, thousands, more, less, many, few, tally, odd, even, every other, skip count, how many times?, multiple of, sequence, continue, predict, pattern, pair, rule, relationship

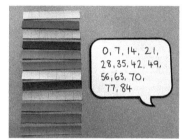

Monday

Give each pupil a sheet of 2cm squared paper and some colouring pencils.

Ask if pupils can think of things that come in sevens (e.g. days of the week). Each pupil colours the squares in strips of 7, then cuts out 12 strips and places them in a named envelope for use throughout the week. (This activity would work equally well with objects grouped into sevens).

Ask pupils to lay out their strips. Then count together from 0–84, pointing to each strip of 7 as they count. Practise counting both forwards and backwards.

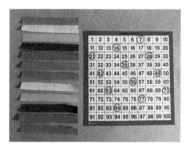

Tuesday

Give each pupil a 100 square and their set of strips.

Repeat Monday's activity. This time, when you say out loud a multiple of 7, pupils circle that number on their 100 square. Count both forwards and backwards.

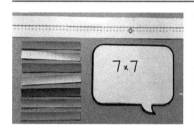

Wednesday

Give each pupil or pair of pupils a number line from 0–100 and their set of strips.

Say out loud a question from the 7 multiplication table and ask pupils to find that multiple of 7 using the strips. They should then circle the total on their number line. Count both forwards and backwards.

Thursday

Give each pupil or pair of pupils 12 sticky notes (or small squares of paper).

Ask pupils to write the multiples of 7 from 7–84 on the notes and then to place them in order from smallest to largest. Count forwards and backwards using the numbers as a resource. Keep the resource for Friday.

Friday

Pupils again lay out their sticky notes in order from smallest to largest. Show a playing card. Pupils count in sevens that number of times and hold up the note that is the multiple shown on the card.

Week 4: Counting

Count from 0 in multiples of 25

Resources: 2cm squared paper, scissors, playing cards

Vocabulary: counting, number, zero, one, two, three …, ten, twenty … one hundred, two hundred … one thousand, how many …?, count, count (up) to / on / back, count in … sixes, sevens, nines, twenty-fives, thousands, more, less, many, few, tally, odd, even, every other, skip count, how many times?, multiple of, sequence, continue, predict, pattern, pair, rule, relationship

Monday

Give each pupil two 10 × 10 grids of squared paper.

Model and ask pupils to copy you cutting the 100 squares in each grid into four groups of 25. Then place these back together into hundreds as shown.

Count from 0–200 and back again in multiples of 25, encouraging a counting rhythm and highlighting the patterns when counting. Keep the resource to use throughout the week.

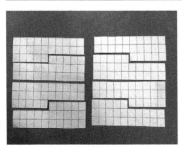

Tuesday

Give each pupil a whiteboard and pen.

Count together from 0–500 and back again in multiples of 25, with pupils recording the numbers on their whiteboards. Model the format as shown. Ask pupils what they notice about the pattern when counting in twenty-fives. Photograph the whiteboards for reference throughout the week.

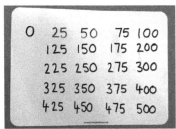

Wednesday

Give each pupil a sheet of large squared paper.

Ask pupils to cut out a grid that is 4 × 10 squares. Each square represents 25. Count forwards together in twenty-fives from 0–1000, with pupils writing in the counting sequence as they chant. Use this to count backwards from 1000–0. (Pupils may wish to keep this resource for Thursday and Friday.)

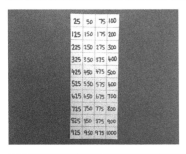

Thursday

Give each pupil a whiteboard and pen.

Emphasise that 4 × 25 = 100. Show a playing card. Pupils count silently in twenty-fives that number of counting steps and write the multiple of 25 on their whiteboards to show you. Repeat.

Friday

Give each pupil a whiteboard and pen.

Emphasise that 4 × 25 = 100. Show a playing card. Pupils *double* that number, and then count silently in twenty-fives that number of counting steps and write the multiple of 25 on their whiteboards to show you. Repeat.

Week 5: Counting

Count from 0 in multiples of 1000

Resources: sticky notes, number lines, counting stick, 0–20 bead string

Vocabulary: counting, number, zero, one, two, three …, ten, twenty … one hundred, two hundred … one thousand, how many …?, count, count (up) to / on / back, count in … sixes, sevens, nines, twenty-fives, thousands, more, less, many, few, tally, odd, even, every other, skip count, how many times?, multiple of, sequence, continue, predict, pattern, pair, rule, relationship

---

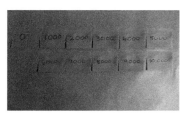

Monday

Give each pupil 11 sticky notes (or small squares of paper).

Model and ask pupils to copy as you write numbers 0–10 000 in multiples of 1000 on the sticky notes. Count from 0–10 000 and back again in multiples of 1000, encouraging a counting rhythm and highlighting the patterns when counting. Keep the resource for use throughout the week.

Tuesday

Give each pair of pupils a number line (or strip of paper or use masking tape on the table top), with 11 evenly spaced marks.

Mark on 0 and 10 000. Then count together from 0–10 000, with pupils marking the number line as they count. Then count backwards using this resource. Repeat to gather fluency.

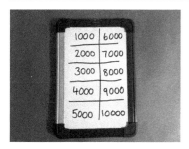

Wednesday

Give each pupil a whiteboard and pen.

Using a counting stick, count together from 0–10 000 in thousands and back.

Pupils draw a 2 × 5 grid on their whiteboards as shown. Ask pupils to write the multiples of 1000 from 1000–10 000. Count from 1000–10 000 and back again, in multiples of 1000, encouraging a counting rhythm.

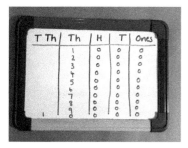

Thursday

Give each pupil a whiteboard and pen.

Pupils draw a place-value grid on their whiteboards as shown. Then they write the counting sequence, from 0–10 000, in thousands. Count forwards and backwards in thousands, using the resource as a prompt.

Ask pupils when they think they will need another place-value column.

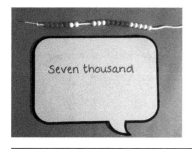

Friday

Give each pupil a whiteboard and pen.

Show a 0–20 bead string. Count from 0–20 and back again. Explain that each bead now has a value of 1000. Count from 0–20 000 and back again. Slide beads across and ask pupils to call out or write the number shown.

Week 6: Counting

Count backwards through 0 to include negative numbers

Resources: counting stick, dice, counters, A4 paper

Vocabulary: counting, number, zero, one, two, three …, ten, twenty … one hundred, two hundred … one thousand, how many …?, count, count (up) to / on / back, count in … sixes, sevens, nines, twenty-fives, thousands, more, less, many, few, tally, odd, even, every other, skip count, how many times?, multiple of, sequence, continue, predict, pattern, pair, rule, relationship

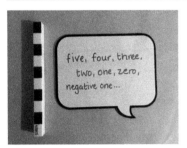

Monday

Ask pupils when they might need to use a negative number (less than zero) in real life (e.g. a thermometer).

Use a counting stick held vertically to count from 5 to –5 and back to 5 again. Repeat.

Tuesday

Give each pupil a whiteboard and pen.

Ask pupils to draw a bead string with 10 beads as shown, with a line drawn to indicate zero. Place the whiteboard vertically and count forwards and backwards from 5 to –5 and back again.

Wednesday

Give each pair of pupils a dice, a counter and a sheet of A4 paper.

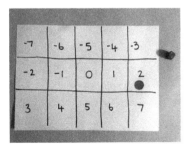

Ask pupils to draw a 5 × 3 grid with numbers as shown. Partner 1 in each pair is 'positive' and partner 2 is 'negative'. The counter is placed on 0. Partner 1 rolls the dice and moves the counter that number of spaces forward along the positive numbers (e.g. rolls 2 and moves to +2). Then partner 2 rolls the dice and moves the counter that number of spaces back towards the negative numbers (e.g. rolls 4 and moves back to –2). Repeat; the winner is the pupil who reaches either 7 or –7 first.

Thursday

Repeat Wednesday's activity, with pupils swapping roles.

Friday

Give each pupil a whiteboard and pen.

On the board, write a range of numbers between –20 and 20. Ask pupils to order the numbers from smallest to largest. Repeat, ordering from largest to smallest. (It may help to write the numbers vertically to mimic the context of a thermometer.)

Week 1: Place value

Recognise the place value of each digit in a four-digit number

Resources: place-value counters

Vocabulary: place value, place, ones, tens, hundreds, thousands, digit, one-, two-, three-, four-digit number, 'teens' number, represents, exchange, the same as, equal to, greater, more, larger, less, fewer, smaller, greatest, most, largest, least, fewest, smallest, one / ten / hundred / thousand more or less, compare, order, first, second, third … last, numeral, consecutive

Monday

Give each pupil some place-value counters and a whiteboard and pen.

Ask pupils to draw a place-value grid on their whiteboards as shown. Call out a range of four-digit numbers and ask pupils to put counters on their grid to represent the number of thousands, hundreds, tens and ones. (Depending on how familiar your pupils are with place-value counters, it may be useful to show the link between them and Dienes.)

Tuesday

Repeat Monday's activity, this time calling out some four-digit numbers that include zeros in the hundreds, tens or ones columns.

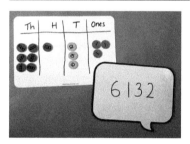

Wednesday

Give each pupil some place-value counters and a whiteboard and pen.

Ask pupils to draw a place-value grid on their whiteboards as shown. Call out a four-digit number. Ask pupils to add one more and represent using the counters. They could then write the number sentence underneath.

Repeat, this time finding one less.

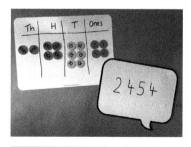

Thursday

Repeat Wednesday's activity, this time asking pupils to add ten more. They could then write the number sentence underneath.

Repeat, this time finding ten less.

Friday

Repeat Wednesday's activity, this time asking pupils to add one hundred more. They could then write the number sentence underneath.

Repeat, this time finding one hundred less.

Week 2: Place value

Order and compare numbers beyond 1000

Resources: place-value counters, number lines, 0–9 digit cards

Vocabulary: place value, place, ones, tens, hundreds, thousands, digit, one-, two-, three-, four-digit number, 'teens' number, represents, exchange, the same as, equal to, greater, more, larger, less, fewer, smaller, greatest, most, largest, least, fewest, smallest, one / ten / hundred / thousand more or less, compare, order, first, second, third … last, numeral, consecutive

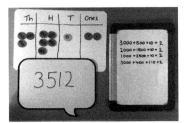

Monday

Give each pupil some place-value counters and a whiteboard and pen.

Ask pupils to draw a place-value grid on their whiteboards as shown. Say a four-digit number. Ask pupils to make the number using place-value counters. Then challenge them to find ten different ways of partitioning that number. Pupils should record this using the expanded form.

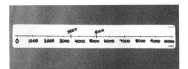

Tuesday

Give each pupil an empty number line (or a strip of paper or use masking tape on the desk tops) with ten intervals.

Ask pupils to mark the thousands from 0–10 000 on their number line. (Highlight the use of spaces every three places when writing numbers in standard form, and explain that sometimes commas are used instead.) Call out a four-digit number and ask pupils to mark that number on the number line. Repeat with other numbers.

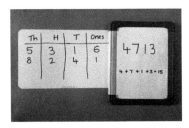

Wednesday

Give each pupil a whiteboard and pen.

Ask pupils to draw a place-value grid on their whiteboards as shown. Write four digits on the board. Ask what the digits add up to. Ask pupils to find as many four-digit numbers with the same digit total as they can. (Digit cards may help with this activity.)

Thursday

Give each pair of pupils a set of digit cards.

Write a four-digit number and either the < or > symbol on the board. Both pupils use their digit cards to rearrange the given digits to create another four-digit number that is greater than or less than the number shown.

Friday

Give each pupil or pair of pupils a set of digit cards.

Write four digits on the board. Ask pupils to use their digit cards to create a number that has specific properties (e.g. smallest odd number using the digits, largest even number). Repeat.

Week 3: Place value

Find 1000 more or less than a given number

Resources: place-value counters

> **Vocabulary:** place value, place, ones, tens, hundreds, thousands, digit, one-, two-, three-, four-digit number, 'teens' number, represents, exchange, the same as, equal to, greater, more, larger, less, fewer, smaller, greatest, most, largest, least, fewest, smallest, one / ten / hundred / thousand more or less, compare, order, first, second, third … last, numeral, consecutive

Monday

Give each pupil some place-value counters and a whiteboard and pen.

Ask pupils to draw a place-value grid on their whiteboards as shown. Remind them how to use counters to add ones, tens, hundreds and thousands and how to record a number sentence. Write a four-digit number on the board. Ask pupils to add one more counter to the thousands column and write the number sentence underneath. Repeat.

Tuesday

Give each pupil some place-value counters and a whiteboard and pen.

Ask pupils to draw a place-value grid on their whiteboards as shown. Write a four-digit number on the board. Ask pupils to show the number using counters and then to remove a thousand counter from the thousands column. They write the number sentence underneath. Repeat.

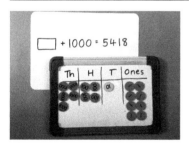

Wednesday

Give each pupil some place-value counters and a whiteboard and pen.

Ask pupils to draw a place-value grid on their whiteboards as shown. Write 5418 on the board and explain that 1000 has already been added to give this answer. Model writing '☐ + 1000 = 5418'. Ask what number goes in the box and ask pupils to represent it using counters. Repeat with other four-digit numbers.

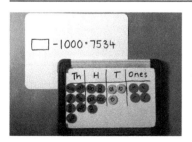

Thursday

Give each pupil some place-value counters and a whiteboard and pen.

Ask pupils to draw a place-value grid on their whiteboards as shown. Write 7534 on the board and explain that 1000 has already been subtracted to give this answer. Model writing '☐ – 1000 = 7534'. Ask what number goes in the box and ask pupils to represent it using counters. Repeat with other four-digit numbers.

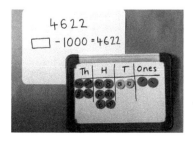

Friday

Give each pair of pupils a whiteboard and pen each.

Partner 1 writes a four-digit number on their whiteboard and says that 1000 has already been added / subtracted to give this answer. Partner 2 represents the calculation using place value counters, and writes the corresponding number sentence. Partner 1 checks the answer. Swap roles and repeat.

Week 4: Place value

Resources: 0–9 digit cards, counters

> **Vocabulary:** place value, place, ones, tens, hundreds, thousands, digit, one-, two-, three-, four-digit number, 'teens' number, represents, exchange, the same as, equal to, greater, more, larger, less, fewer, smaller, greatest, most, largest, least, fewest, smallest, one / ten / hundred / thousand more or less, compare, order, first, second, third … last, numeral, consecutive

Monday

Give each pair of pupils a set of 0–9 digit cards.

Pairs lay out four cards to make a number with different properties as directed by the teacher (e.g. a four-digit number with consecutive digits, a four-digit number which is even). Pairs look at the remaining six cards. Partner 1 in each pair finds the largest four-digit number and partner 2 finds the smallest four-digit number that can be made. Pupils check each other's work. Repeat.

Tuesday

Repeat Monday's activity, with pupils swapping roles.

Wednesday

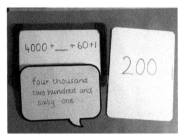

Give each pair of pupils 15 counters and a whiteboard and pen.

Tell pupils that you have used 15 counters to make a four-digit number on your place-value grid. Pairs try to work out the number you could have made. They could record their answers using numbers, expanded form or drawings.

Thursday

Give each pupil a whiteboard and pen.

Write a four-digit number in expanded form on the board with either the thousands, hundreds, tens or ones value missing as shown. Say the number and ask pupils to write the missing value on their whiteboards to show you. Repeat.

Friday

Give each pupil a whiteboard and pen.

Explain that a palindrome is a number, word or sentence that reads the same forwards or backwards. Write 2662 as an example and ask pupils to read out the number. Ask pupils to write a four-digit palindrome of their own with varying properties (e.g. includes the digits 4 and 7, has a digit total of 15, is even).

Week 5: Place value

Solve number problems involving place value ideas with increasingly large positive numbers

Resources: no additional resources required

Vocabulary: place value, place, ones, tens, hundreds, thousands, digit, one-, two-, three-, four-digit number, 'teens' number, represents, exchange, the same as, equal to, greater, more, larger, less, fewer, smaller, greatest, most, largest, least, fewest, smallest, one / ten / hundred / thousand more or less, compare, order, first, second, third … last, numeral, consecutive

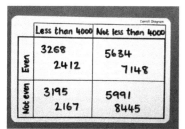

Monday

Give each pupil or pair of pupils a whiteboard and pen.

Ask pupils how many four-digit numbers they can create where the digits are consecutive. Can they create a four-digit number whose digits are consecutive and which is even, odd, a multiple of 5, a multiple of 10, etc.?

Tuesday

Give each pupil or pair of pupils a whiteboard and pen.

Ask pupils to draw a Carroll diagram labelled 'Less than 4000', 'Not less than 4000', 'Even' and 'Not Even'. Call out four-digit numbers and ask pupils to write them on the Carroll diagram. Finally, ask if they can place a four-digit number of their own in each quadrant.

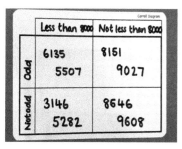

Wednesday

Repeat Tuesday's activity, but this time label the Carroll diagram 'Less than 8000', 'Not less than 8000', 'Odd' and 'Not odd'.

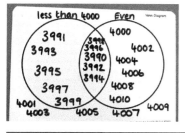

Thursday

Give each pupil or pair of pupils a whiteboard and pen.

Ask pupils to draw a Venn diagram labelled 'Less than 4000' and 'Even'. Ask pupils to write all the numbers between 3990 and 4010 in the correct places on the Venn diagram.

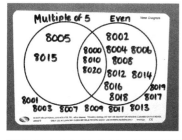

Friday

Give each pupil or pair of pupils a whiteboard and pen.

Ask pupils to draw a Venn diagram labelled 'Multiple of 5' and 'Even'. Ask pupils to write all the numbers between 8000 and 8020 on the Venn diagram.

Repeat with other ranges of numbers.

Solve practical problems involving place value ideas with negative numbers

Resources: dice

Vocabulary: place value, place, ones, tens, hundreds, thousands, digit, one-, two-, three-, four-digit number, 'teens' number, represents, exchange, the same as, equal to, greater, more, larger, less, fewer, smaller, greatest, most, largest, least, fewest, smallest, one / ten / hundred / thousand more or less, compare, order, first, second, third … last, numeral, consecutive

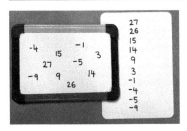

Monday

Give each pupil a whiteboard and pen.

Write ten numbers on the board with values between −10 and 30. Explain that these are temperatures taken over ten days. Ask pupils to order the numbers from smallest to greatest. Repeat.

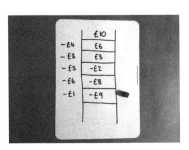

Tuesday

Give each pair of pupils a dice and a whiteboard and pen each.

Ask pupils to work in pairs. Each pupil draws a 'ladder' with five spaces, as shown, and writes £10 at the top. Pupils take turns to roll their dice and deduct the amount shown. Pupils deduct five amounts in the same way. This represents pocket money of £10 being spent each week. The pupil with the least debt after five rolls wins.

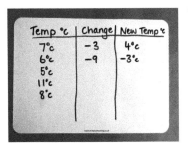

Wednesday

Give each pupil a whiteboard and a pen.

Ask pupils to draw a grid as shown. Tell them that the temperature one morning was −2°C. Explain that the temperature had dropped overnight but that it had been a single-digit positive temperature to begin with. What were the possible start temperatures and what was the decrease in temperature for each?

Thursday

Give each pupil a whiteboard and pen.

Ask pupils to draw a grid as shown. Call out a number which pupils write in the 'Temp °C' column. Next, say a single-digit temperature change in the range −9 to 9. Pupils write this in the middle column. They then calculate the new temperature and write it in the third column. (It may be useful to display a thermometer or a vertical −20 to 20 number line.)

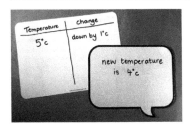

Friday

Give each pupil a whiteboard and pen.

Ask pupils to draw a grid as shown. Say a temperature value between −20 and 20. Pupils write the number in the 'Temperature' column. Next, tell pupils a new temperature (saying the temperature in words). Ask pupils to write in / describe the temperature change. Repeat several times.

Week 1: Representing numbers

Identify, represent and estimate numbers using different representations

Resources: Dienes

Vocabulary: place value, place, ones, tens, hundreds, thousands, digit, one-, two-, three- or four-digit number, 'teens' number, represents, exchange, the same as, equal to, greater, more, larger, less, fewer, smaller, greatest, most, largest, least, fewest, smallest, one / ten / hundred / thousand more or less, compare, order, first, second, third … last, numeral, consecutive, estimate, nearly, roughly, close to, approximate, exactly, too many / few, round up / down / to, nearest, Roman numerals (I, V, X, L, C)

Monday

Give each pupil a whiteboard and pen.

Show a four-digit number using Dienes. Pupils should write the number on their whiteboards to show you. Repeat with other numbers. This activity could be linked to estimating, with pupils estimating quantities using Dienes, bowls of pasta / cereal or small natural objects such as acorns, leaves or petals.

Tuesday

Give each pupil a whiteboard and pen.

Write a four-digit number in partitioned form on the board. Pupils should represent the number on a place-value grid as shown. Repeat with other numbers.

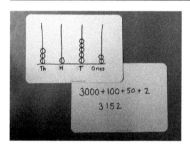

Wednesday

Give each pupil a whiteboard and pen.

On a whiteboard, draw a representation of a four-digit number using an abacus model. Ask pupils to write the four-digit number in partitioned form and then say the number together. Repeat with other numbers.

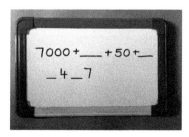

Thursday

Give each pupil a whiteboard and pen.

Write a four-digit number in partitioned and written form on the board. Explain that some values are missing. Pupils should write the four-digit number in numerals, show you and then say it together. Repeat with other numbers.

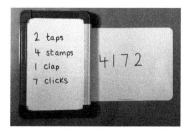

Friday

Give each pupil a whiteboard and pen.

Explain that a stamp of the foot represents thousands, claps represent hundreds, clicks of the fingers represent tens and a tap on the knees represents ones. Make a number using this body percussion. Pupils write the number on their whiteboards, show you and say the number out loud. Repeat with other numbers. (The sequence does not always have to be in place-value order, as shown here.)

Week 2: Representing numbers

Resources: lolly sticks

Vocabulary: place value, place, ones, tens, hundreds, thousands, digit, one-, two-, three- or four-digit number, 'teens' number, represents, exchange, the same as, equal to, greater, more, larger, less, fewer, smaller, greatest, most, largest, least, fewest, smallest, one / ten / hundred / thousand more or less, compare, order, first, second, third … last, numeral, consecutive, estimate, nearly, roughly, close to, approximate, exactly, too many / few, round up / down / to, nearest, Roman numerals (I, V, X, L, C)

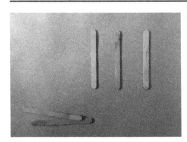

Monday

Give each pupil five lolly sticks.

Ask pupils for examples of where they have seen Roman numerals (e.g. clock faces, lists, film titles).

Introduce the Roman numerals I and V, and then model the Roman numerals for the numbers I, 2, 3, 5, 6, 7, 8. Pupils create each numeral in turn using lolly sticks.

Tuesday

Give each pupil five lolly sticks and a whiteboard and pen.

Recap Monday's activity by calling out numbers and asking pupils to make the corresponding Roman numerals.

Now tell pupils they can only use up to three lolly sticks. Which Roman numerals from I, 2, 3, 5, 6, 7, 8 can they make?

Wednesday

Give each pupil a whiteboard and pen.

Explain that the digits in Roman numerals are arranged from largest to smallest value (e.g. VIII), except when I, X or C is placed in front of a numeral for subtractive notation, and explain that the Romans did not repeat more than three of the same digit in a row. Ask pupils if VIIII for 9 would be possible given this information (no). Explain that X represents 10 and 9 is written IX (10 – 1). Call out numbers and ask pupils to write the corresponding Roman numerals.

Thursday

Give each pupil a whiteboard and pen.

Ask pupils to draw a 3 × 2 bingo board and to write any six numbers between 1 and 10. Write a Roman numeral on the board. If pupils have that number on their bingo board, they cross it off. Repeat until someone shouts 'Bingo!'.

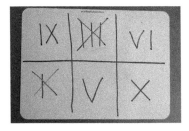

Friday

Give each pupil a whiteboard and pen.

Ask pupils to draw a 3 × 2 bingo board and to write any six Roman numerals between 1 and 10. Write a number on the board. If pupils have that number on their bingo board, they cross it off. Repeat until someone shouts 'Bingo!'.

Week 3: Representing numbers

Read Roman numerals to 50

Resources: lolly sticks

Vocabulary: place value, place, ones, tens, hundreds, thousands, digit, one-, two-, three- or four-digit number, 'teens' number, represents, exchange, the same as, equal to, greater, more, larger, less, fewer, smaller, greatest, most, largest, least, fewest, smallest, one / ten / hundred / thousand more or less, compare, order, first, second, third … last, numeral, consecutive, estimate, nearly, roughly, close to, approximate, exactly, too many / few, round up / down / to, nearest, Roman numerals (I, V, X, L, C)

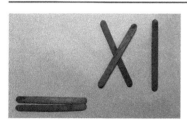

Monday

Give each pupil five lolly sticks.

Recap Roman numerals to 10 by repeating the Week 2 Friday activity.

Introduce the Roman numerals for the numbers 10–13. Pupils create each numeral in turn using lolly sticks.

Tuesday

Give each pupil five lolly sticks.

Recap Monday's activity by calling out numbers and asking pupils to use lolly sticks to make the corresponding Roman numerals. Extend to include XIV (14) and XV (15), making links to our place-value system. Tell pupils they can use exactly five lolly sticks. Which Roman numerals between 1 and 15 can they make (VIII, XIII and XIV)?

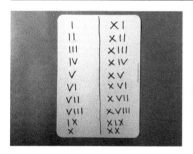

Wednesday

Give each pupil a whiteboard and pen.

Remind pupils that the digits in Roman numerals are arranged from largest to smallest value, apart from when using subtractive notation, and that the Romans did not repeat more than three of the same numeral in a row. Ask pupils to create two columns on their whiteboards. In one column they write the Roman numerals 1–10 and in the other the Roman numerals 11–20. What is similar?

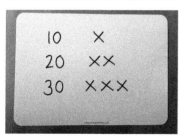

Thursday

Give each pupil a whiteboard and pen.

Ask pupils to write 10, 20 and 30 as Roman numerals.

Remind pupils that the Romans did not repeat more than three of the same numerals in a row. Then ask if XXXX represents 40. Why not? Introduce L and the numerals for 40 (XL) and 50 (L).

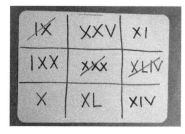

Friday

Give each pupil a whiteboard and pen.

Ask them to draw a 3 × 3 bingo board and to write any nine Roman numerals between 1 and 50. Write a number on the board. If pupils have that number on their bingo board, they cross it off. Repeat until someone shouts 'Bingo!'.

Week 4: Representing numbers

Read Roman numerals to 100

Resources: no additional resources required

> **Vocabulary:** place value, place, ones, tens, hundreds, thousands, digit, one-, two-, three- or four-digit number, 'teens' number, represents, exchange, the same as, equal to, greater, more, larger, less, fewer, smaller, greatest, most, largest, least, fewest, smallest, one / ten / hundred / thousand more or less, compare, order, first, second, third … last, numeral, consecutive, estimate, nearly, roughly, close to, approximate, exactly, too many / few, round up / down / to, nearest, Roman numerals (I, V, X, L, C)

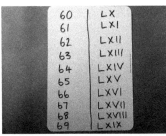

Monday

Give each pupil a whiteboard and pen.

Recap Roman numerals to 50 by repeating the Week 3 Friday activity.

Introduce C as the numeral for 100. Together, pupils count in tens from 10–100, writing the numbers on their whiteboards in the first column and the corresponding Roman numerals in the second column.

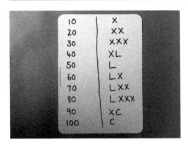

Tuesday

Give each pupil a whiteboard and pen.

Ask pupils to write the numbers 60–69 in the left-hand column. Model writing 60–69 in Roman numerals and ask pupils to copy them into the right-hand column of their whiteboards.

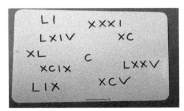

Wednesday

Repeat Tuesday's activity for the numbers 70–79, 80–89 and 90–99. Recap 100.

Ask: what is similar when writing the numerals? What is different? Clarify that the ones value is represented in the same way, only the tens value changes.

Thursday

Give each pupil a whiteboard and pen.

Call out ten numbers and ask pupils to write them in Roman numerals. Then ask pupils to write them in order from smallest to largest.

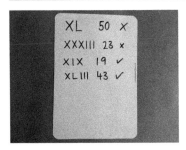

Friday

Give each pupil a whiteboard and pen.

This is an opportunity for pupils to spot errors and check for accuracy. Write a Roman numeral from 1–100 on the board for pupils to copy, and then say the number in words. If you say it correctly, pupils tick the number they have written; otherwise they give it a cross. Repeat with other numbers.

Week 5: Representing numbers

Round any number to the nearest 10 or 100

Resources: number lines / counting sticks, 0–9 digit cards

Vocabulary: place value, place, ones, tens, hundreds, thousands, digit, one-, two-, three- or four-digit number, 'teens' number, represents, exchange, the same as, equal to, greater, more, larger, less, fewer, smaller, greatest, most, largest, least, fewest, smallest, one / ten / hundred / thousand more or less, compare, order, first, second, third … last, numeral, consecutive, estimate, nearly, roughly, close to, approximate, exactly, too many / few, round up / down / to, nearest

---

Monday

Give each pupil an empty number line (or mini counting stick) with ten intervals marked on it.

Ask pupils to mark 60 and 70 on the number line. Call out a number (e.g. 61) and ask pupils to mark it on the line. Then ask whether 60 or 70 is the nearest multiple of 10. Repeat with other numbers. Explain that this is rounding to the nearest ten. Highlight that 5 is always rounded up.

Tuesday

Give each pupil an empty number line (or mini counting stick) with ten intervals marked on it.

Ask pupils to mark 340 and 350 on the number line. Call out a number (e.g. 343) and ask pupils to mark it on the line. Then ask whether 340 or 350 is the nearest multiple of 10. Repeat with other numbers. Explain that this is rounding to the nearest ten in a three-digit number.

Wednesday

Give each pupil an empty number line (or mini counting stick) with ten intervals marked on it.

Ask pupils to mark 200 to 300 in tens on the number line. Call out a number (e.g. 270) and ask pupils to mark it on the line. Then ask whether 200 or 300 is the nearest multiple of 100. Repeat with other numbers. Explain that this is rounding to the nearest hundred. Highlight that 50 is always rounded up.

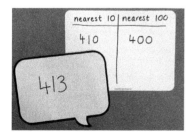

Thursday

Give each pupil a whiteboard and pen.

Ask pupils to split their board in half as shown. Call out a number. Ask pupils to round it to the nearest ten and then to the nearest hundred.

Friday

Give each pair of pupils a set of 0–9 digit cards and a whiteboard and pen.

Partner 1 creates a three-digit number and says the number. Partner 2 rounds it to the nearest hundred. If they answer correctly, they score a point. Swap roles and repeat. Who scores most points?

Week 6: Representing numbers

Round any number to the nearest 10, 100 or 1000

Resources: number lines / counting sticks

> **Vocabulary:** place value, place, ones, tens, hundreds, thousands, digit, one-, two-, three- or four-digit number, 'teens' number, represents, exchange, the same as, equal to, greater, more, larger, less, fewer, smaller, greatest, most, largest, least, fewest, smallest, one / ten / hundred / thousand more or less, compare, order, first, second, third … last, numeral, consecutive, estimate, nearly, roughly, close to, approximate, exactly, too many / few, round up / down / to, nearest

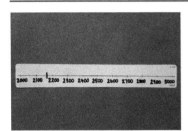

Monday

Give each pupil an empty number line (or mini counting stick) with ten intervals marked on it.

Ask pupils to mark 2000 and 3000 on the number line. Call out a number (e.g. 2173) and ask pupils to mark it on the line. Then ask whether 2000 or 3000 is the nearest multiple of 1000. Repeat with other numbers. Explain that this is rounding to the nearest thousand. Highlight that 500 is equidistant to the nearest hundred and is always rounded up.

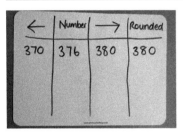

Tuesday

Give each pupil a whiteboard and pen.

Ask pupils to split their board in four and to label it as shown. (← and → indicate the previous and next multiples of 10.) Write 376 and ask pupils to round it to the nearest ten. Model completing the grid. Repeat with other numbers.

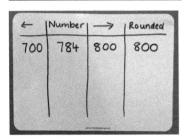

Wednesday

Give each pupil a whiteboard and pen.

Ask pupils to split their board in four and to label it as shown. (← and → indicate the previous and next multiples of 100.) Write 784 and ask pupils to round it to the nearest hundred. Model completing the grid. Repeat with other numbers.

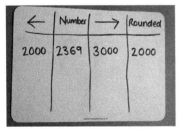

Thursday

Give each pupil a whiteboard and pen.

Ask pupils to split their board in four and to label it as shown. (← and → indicate the previous and next multiples of 1000.) Write 2369 and ask pupils to round it to the nearest thousand. Model completing the grid. Repeat with other numbers.

Friday

Give each pupil a whiteboard and pen.

Tell pupils you have rounded a number to the nearest ten and the answer is 260. What are the possible numbers? (Highlight that 260, when rounded to the nearest ten, would be 260, i.e. multiples of 10 round to themselves.) Repeat with other numbers.

Week 1: Addition and subtraction

Add mentally a four-digit number and ones

Resources: dice

Vocabulary: digit, add, addition, more / less, plus, make, sum, total, altogether, one more, two more … ten more, one hundred more … one thousand more, how many more to make …?, missing number, how many / much more is …?, subtract, subtraction, take away, minus, leave, how many are left / left over?, one less, two less … ten less … one hundred less … one thousand less, how many fewer?, how much less?, difference between, +, −, =, equals, sign, is the same as, boundary, exchange

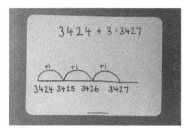

Monday

Give each pair of pupils a whiteboard and pen.

Write 3424 + 3 on the board. Ask pairs to discuss how they will solve this. Record their ideas on the board to display throughout the week.

Model drawing a number line and counting on 3 ones from 3424. Repeat with other four-digit numbers, adding ones (but not crossing the tens boundary).

Tuesday

Give each pair of pupils a dice and a whiteboard and pen.

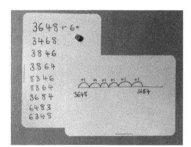

Start by modelling the number line (counting on) method. Pairs either follow this model using their whiteboards or use 'make 10' strategies mentally.

Write 3648 on the board. Ask pupils to arrange the digits in at least ten ways to generate different four-digit numbers. Partner 1 in each pair rolls the dice and adds the number shown on the dice to each four-digit number. Partner 2 checks the answers by counting along the number line. Encourage using known facts (e.g. I know that 6 + 3 = 9, therefore 3846 + 3 = 3849).

Wednesday

Repeat Tuesday's activity, with pupils swapping roles.

Thursday

Give each pupil a whiteboard and pen.

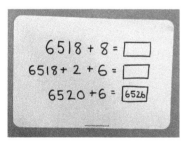

Write 6518 + 8 on the board and say the number sentence out loud together. Explain that, when we cross the tens boundary, it can be useful to make links. Remind pupils of the 'make 10' strategy used in Tuesday's activity. So, by partitioning 8, we can make the question 6518 + 2 + 6 = 6520 + 6. Allow pupils time to practise, in pairs if wished, with calculations that cross the tens boundary (e.g. 2723 + 8, 6547 + 6, 9325 + 7, 5523 + 9).

Friday

Write 4738 on the board and explain that 6 has already been added to give this answer. Write '☐ + 6 = 4738' on the board. Ask what number goes in the box, challenging pupils to explain how they calculate it (either verbally or in writing).

Repeat with other calculations in the same format.

Week 2: Addition and subtraction

Subtract mentally a four-digit number and ones

Resources: dice

> **Vocabulary:** digit, add, addition, more / less, plus, make, sum, total, altogether, one more, two more … ten more, one hundred more … one thousand more, how many more to make …?, missing number, how many / much more is …?, subtract, subtraction, take away, minus, leave, how many are left / left over?, one less, two less … ten less … one hundred less … one thousand less, how many fewer?, how much less?, difference between, +, –, =, equals, sign, is the same as, boundary, exchange

Monday

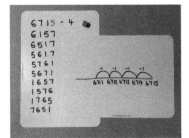

Write 6819 – 6 on the board. Ask pupils to discuss how they will solve this. Record their ideas on the board to display throughout the week.

Model drawing a number line and counting back 6 ones from 6819. Repeat with other four-digit numbers, subtracting a one-digit number (but not crossing the tens boundary).

Tuesday

Give each pair of pupils a dice and a whiteboard and pen.

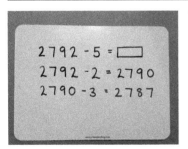

Start by modelling the number line (counting back) method. Pairs either follow this model using their whiteboards or use 'make 10' strategies mentally.

Write a four-digit number on the board. Ask pupils to arrange the digits at least ten ways to generate different four-digit numbers. Partner 1 in each pair rolls the dice and subtracts the number shown on the dice from each four-digit number. Partner 2 checks the answers by counting back along the number line.

Wednesday

Repeat Tuesday's activity, with pupils swapping roles.

Thursday

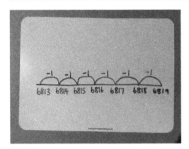

Give each pair of pupils a whiteboard and pen.

Write 2792 – 5 on the board and say the number sentence out loud together. Explain that, when we cross the tens boundary, it can be useful to make links. So, by partitioning 5 into 2 + 3, we can make the question (2792 – 2) – 3 = 2787. Allow pairs time to practise with calculations that cross the tens boundary (e.g. 6223 – 7, 2657 – 9, 4325 – 8, 2423 – 6).

Friday

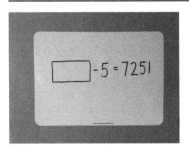

Write 7251 on the board and explain that 5 has already been subtracted to give this answer. Write '☐ – 5 = 7251' on the board. Ask what number goes in the box, challenging pupils to explain how they calculate it (either verbally or in writing).

Repeat with other calculations in the same format.

Week 3: Addition and subtraction

Add mentally a four-digit number and tens / hundreds

Resources: place-value counters / counters / Dienes, cubes

Vocabulary: digit, add, addition, more / less, plus, make, sum, total, altogether, one more, two more … ten more, one hundred more … one thousand more, how many more to make …?, missing number, how many / much more is …?, subtract, subtraction, take away, minus, leave, how many are left / left over?, one less, two less … ten less … one hundred less … one thousand less, how many fewer?, how much less?, difference between, +, −, =, equals, sign, is the same as, boundary, exchange

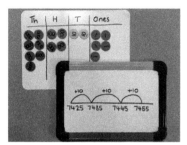

Monday

Give each pair of pupils some place-value counters (or plain counters or Dienes).

Write 7425 + 30 on the board. Ask pupils to discuss how they will solve this. Record their ideas on the board to display throughout the week.

Model drawing a place-value grid on the board and adding 3 tens. Then draw a number line to represent the same calculation. Repeat with 3728 + 50 and 1665 + 40.

Tuesday

Give each pupil 10 cubes and a whiteboard and pen.

Ask pupils to make their own 10-cube counting stick to use throughout the week. Write 3182 on the board and all say the number together. Count forwards and backwards in tens from 3182, noting the counting steps. Repeat with other start numbers. Ask pupils which digit(s) change each time.

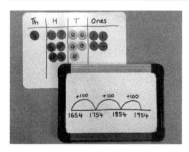

Wednesday

Give each pair of pupils a handful of place-value counters and a whiteboard and pen.

Write 1654 + 300 on the board. Ask pupils to discuss how they will solve this. Record their ideas on the board to display throughout the week. Model drawing a place-value grid on the board and adding 3 hundreds using place-value counters (or Dienes or counters). Then draw a number line to represent the same calculation. Repeat with other calculations (e.g. 3428 + 500, 1665 + 200).

Thursday

Give each pair of pupils a counting stick and a whiteboard and pen.

Tell pupils that each cube represents 100. Write 1348 + 400 on the board. Break off 4 cubes and ask how many hundreds are represented (4). Ask the value of the cubes (400). Add 400 to 1348, adding 100 each time. Repeat with other four-digit numbers.

Partner 1 in each pair generates a four-digit number and writes it on their whiteboard. Partner 2 shows a quantity from their stick of 10 for adding. Partner 1 writes the series of calculations and partner 2 checks the answers.

Friday

Repeat Thursday's activity, with pupils swapping roles.

Week 4: Addition and subtraction

Subtract mentally a four-digit number and tens / hundreds

Resources: place-value counters / counters / Dienes, cubes

Vocabulary: digit, add, addition, more / less, plus, make, sum, total, altogether, one more, two more ... ten more, one hundred more ... one thousand more, how many more to make ...?, missing number, how many / much more is ...?, subtract, subtraction, take away, minus, leave, how many are left / left over?, one less, two less ... ten less ... one hundred less ... one thousand less, how many fewer?, how much less?, difference between, +, –, =, equals, sign, is the same as, boundary, exchange

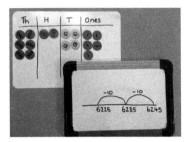

Monday

Give each pair of pupils some place-value counters (or plain counters or Dienes).

Write 6245 – 20 on the board. Ask pupils to discuss how they will solve this. Record their ideas on the board to display throughout the week.

Model drawing a place-value grid on the board and subtracting 2 tens. Then draw a number line to represent the same calculation. Repeat with 5798 – 50 and 2665 – 40.

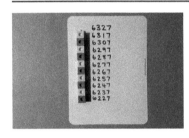

Tuesday

Give each pupil 10 cubes and a whiteboard and pen.

Ask pupils to make their own 10-cube counting stick for use throughout the week. Write 6327 on the board and all say the number together. Count forwards and backwards in tens from 6327, noting the counting steps. Repeat with other start numbers. Ask pupils which digit(s) change each time.

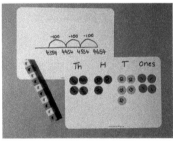

Wednesday

Give each pair of pupils a counting stick and a whiteboard and pen.

Write 4654 – 300 on the board. Ask pupils to discuss how they will solve this. Record their ideas on the board to display throughout the week. Model several strategies: place-value grid and subtracting 3 hundreds, using place-value counters or Dienes; number line, counting backwards. Repeat with other calculations (e.g. 3928 – 400, 1765 – 500).

Thursday

Give each pair of pupils a counting stick and a whiteboard and pen.

Tell pupils that each cube represents 100. Write 6506 on the board. Break off 3 cubes and ask how many hundreds are represented (3). Ask the value of the cubes (300). Subtract 300 from 6506. Count together, pointing at the cubes. Write '6506 – 300 = 6206' on the board.

Partner 1 in each pair generates a four-digit number and writes it on their whiteboard. Partner 2 shows a quantity from their stick of 10 for subtracting. Partner 1 counts backwards to find the answer, using the counting stick for support. Partner 2 checks the answer.

Friday

Repeat Thursday's activity, with pupils swapping roles.

Add mentally a four-digit number and thousands (finding 1000 more)

Resources: counters, tin, cubes, bowls

Vocabulary: digit, add, addition, more / less, plus, make, sum, total, altogether, one more, two more … ten more, one hundred more … one thousand more, how many more to make …?, missing number, how many / much more is …?, subtract, subtraction, take away, minus, leave, how many are left / left over?, one less, two less … ten less … one hundred less … one thousand less, how many fewer?, how much less?, difference between, +, –, =, equals, sign, is the same as, boundary, exchange

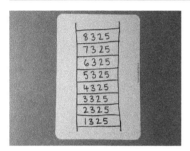

Monday

Give each pair of pupils a whiteboard and pen.

Ask pupils to draw a 'ladder' with eight spaces. In the bottom space, write a four-digit number that has a 1 in the thousands column. Count up in thousands, recording the numbers in each space. When you have finished, count forwards and backwards from that number. Repeat with other start numbers under 1999.

Tuesday

Give each pupil a whiteboard and pen.

Write 2618 + 3000 on the board. Ask pupils to discuss how they will solve this. Start with 2618, which the whole class says together. Explain that, each time you drop a counter into a tin, pupils need to add 1000. Ask again how many thousands are being added (3). Repeat with other calculations (but not crossing the tens boundary).

Wednesday

Give each pair of pupils 10 cubes and a bowl.

Write 2364 + 6000 on the board. Pupils add 1000 to 2364, placing a cube in the bowl to represent adding 1000. Partner 1 in each pair counts out loud to add 6000 and partner 2 checks the number of cubes in the bowl. Repeat with other calculations where a multiple of 1000 is added to a four-digit number.

Thursday

Repeat Wednesday's activity, with pupils swapping roles.

Friday

Write 7199 on the board and explain that 5000 has already been added to give this answer. Write '5000 + ☐ = 7199' on the board. Ask what number goes in the box. Repeat with other calculations in the same format. Remind pupils to note which of the columns changes as we count in thousands – a good check for accuracy!

Week 6: Addition and subtraction

Subtract mentally a four-digit number and thousands (finding 1000 less)

Resources: cubes, place-value counters

Vocabulary: digit, add, addition, more / less, plus, make, sum, total, altogether, one more, two more … ten more, one hundred more … one thousand more, how many more to make …?, missing number, how many / much more is …?, subtract, subtraction, take away, minus, leave, how many are left / left over?, one less, two less … ten less … one hundred less … one thousand less, how many fewer?, how much less?, difference between, +, −, =, equals, sign, is the same as, boundary, exchange

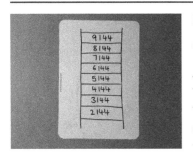

Monday

Give each pair of pupils a whiteboard and pen.

Ask pupils to draw a 'ladder' with eight spaces. In the top space write a four-digit number that has a 9 in the thousands column. Count back in thousands, recording the numbers in each space. When you have finished, count forwards and backwards in thousands from that number, explaining that this is finding 1000 more / less. Repeat with other start numbers over 9000.

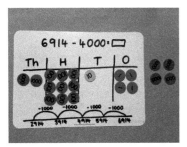

Tuesday

Give each pair of pupils some cubes and a whiteboard and pen.

Write 6914 − 4000 on the board. Ask pupils to discuss how they will solve this, including linking known facts. Record their ideas to display throughout the week.

Model drawing a place-value grid, and subtracting 4 thousands using place-value counters (or Dienes or cubes). Then draw a number line to represent the same calculation. Repeat.

Wednesday

Give each pupil 5 cubes and a whiteboard and pen.

Pupils make their own 5-cube counting stick for use throughout the week. Write 8038 on the board. Say the number together. Count backwards and forwards in thousands from 8038. Repeat with other start numbers over 5000.

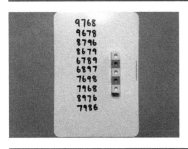

Thursday

Give each pair of pupils a 5-cube counting stick and a whiteboard and pen.

Write 9768 on the board. Pupils use the digits to generate at least 10 four-digit numbers on their whiteboards. Together, they use the counting stick to count forwards and backwards in thousands for each number.

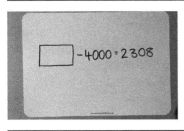

Friday

Write 2308 on the board. Explain that 4000 has already been subtracted to give this answer. Write '☐ − 4000 = 2308' on the board. Ask what number goes in the box. Repeat with other calculations using the same format.

Week 1: Multiplication and division

Recall and use multiplication and division facts for the 6 multiplication table

Resources: counting stick / 100 square, 2cm squared paper, coloured pens, scissors, objects, cubes

Vocabulary: digit, repeated addition, more / less, plus, make, sum, total, altogether, one … ten … one hundred … one thousand more / less, lots, multiplication, groups of, times, multiply, multiplication, multiplied by, multiple of, product, once, twice, three times … ten times as (big, long, wide, etc.), repeated addition, array, row, column, double, halve, share, share equally, one each, two each, three each …, group in pairs, threes … twelves, equal groups of, divide, division, divided by / into, left, left over, remainder, ÷, ×, =, commutative law, commutativity

Monday

Throughout this week, give pupils cube towers, strips of coloured squares (see 'Counting'), small cubes or objects bundled in sixes.

Start by counting in sixes from 0–72 and back again, using a counting stick or a 100 square to support.

Write 'If 6 × 3 = 18, then 3 × 3 = 9' on the board. Ask pupils to prove this in as many ways as they can (e.g. using pictures, number sentences, arrays, groups, objects or number lines for repeated addition). Take photographs to display throughout the week.

Tuesday

Start by counting in threes from 0–72, circling numbers on an individual or interactive 100 square. Then count in sixes from 0–72 and back again, using a different coloured pen on the same 100 square to highlight the relationship between the 3 and the 6 multiplication tables.

Next count together in sixes, forwards and backwards, putting down a tower of 6 cubes each time. Practise again, starting from different multiples of 6.

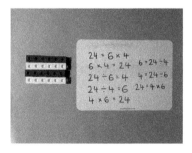

Wednesday

Give each pair of pupils 4 towers of 6 cubes and a whiteboard and pen.

Start by counting in sixes from 0–72 and back again, using either a counting stick or a 100 square to support.

Ask pupils to create an array from their towers and to write as many facts as they can. Now ask how many rows of 6 there are in 24. Model writing '24 ÷ 6 = 4'. Next ask pupils to rotate the array to see how many rows of 4 there are in 24.

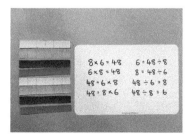

Thursday

Give each pair of pupils 12 strips of 6 squares and a whiteboard and pen.

Call out a multiple of 6 as a multiplication table fact (e.g. 8 × 6 = 48). Ask partner 1 in each pair to make an array that demonstrates this. Partner 2 records all the multiplication and division facts relating to the array.

Friday

Repeat Thursday's activity, with pupils swapping roles.

Recall and use multiplication and division facts for the 9 multiplication table

Resources: objects, cubes, 100 squares

Vocabulary: digit, repeated addition, more / less, plus, make, sum, total, altogether, one ... ten ... one hundred ... one thousand more / less, lots, multiplication, groups of, times, multiply, multiplication, multiplied by, multiple of, product, once, twice, three times ... ten times as (big, long, wide, etc.), repeated addition, array, row, column, double, halve, share, share equally, one each, two each, three each ..., group in pairs, threes ... twelves, equal groups of, divide, division, divided by / into, left, left over, remainder, ÷, ×, =, commutative law, commutativity

Throughout this week, give pupils cube towers, strips of coloured squares (see 'Counting'), small cubes or objects bundled in nines.

Monday

Start by counting in nines from 0–108 and back again using a counting stick or hundred square to support.

Write '9 groups of 5 equals 45' on the board. Ask pupils to prove it in as many ways as they can (e.g. using pictures, number sentences, arrays, groups, objects or number lines for repeated addition). Take photographs to display throughout the week.

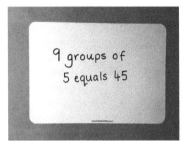

Tuesday

Give each pupil a stick of 9 cubes.

Start by counting in nines from 0–108 and back again, using a counting stick to support.

Count together in nines, forwards and backwards, using the cubes as a resource. (The teacher notes the multiples of 9 on the board or a 100 square.) Practise again, starting from different multiples of 9.

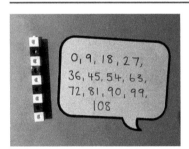

Wednesday

Give each pair of pupils a 100 square.

Ask pupils to circle all the multiples of 9.

Write 'The digits in the products of the 9 multiplication table always add up to 9. True or false? Prove it!' on the board. Pupils work together to prove the statement is true, using all multiples of 9 within the 100 square. Discuss with pupils how the digits of 99 add up to 18, but that the digits of 18 then add to 9.

Thursday

Give each pair of pupils 12 strips of 9 squares and a whiteboard and pen.

Call out a multiple of 9 as a multiplication table fact (e.g. 9 × 4 = 36). Ask partner 1 in each pair to show the correct number of strips of 9 squares to represent this. Partner 2 records all the multiplication and division facts relating to the number fact.

Friday

Repeat Thursday's activity, with pupils swapping roles.

Week 3: Multiplication and division

Resources: objects, Dienes, counting stick

Vocabulary: digit, repeated addition, more / less, plus, make, sum, total, altogether, one ... ten ... one hundred ... one thousand more / less, lots, multiplication, groups of, times, multiply, multiplication, multiplied by, multiple of, product, once, twice, three times ... ten times as (big, long, wide, etc.), repeated addition, array, row, column, double, halve, share, share equally, one each, two each, three each ..., group in pairs, threes ... twelves, equal groups of, divide, division, divided by / into, left, left over, remainder, ÷, ×, =, commutative law, commutativity

Throughout this week, Dienes can be useful for making the clear link that 11 equals 10 + 1; alternatively, give each pupil cube towers, strips of coloured squares (see 'Counting'), small cubes or objects bundled in elevens.

Monday

Start by counting in elevens from 0–132 and back again. (Note that using a 100 square to support limits the multiples to within 9 × 11.)

Write '7 groups of 11 equals 77' on the board. Ask pupils to prove it in as many ways as they can (e.g. using pictures, number sentences, arrays, groups, objects or number lines for repeated addition). Take photographs to display throughout the week.

Tuesday

Start by counting in elevens from 0–132 and back again. Use Dienes to support the mental image of what 11 'looks like'.

Count together in elevens, forwards and backwards using the resource.

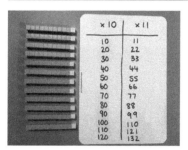

Wednesday

Give each pupil a whiteboard and pen.

Start by counting in elevens from 0–132 and back again, using a counting stick to support.

Pupils divide their whiteboards and label the columns ×10 and ×11 as shown. Together, write all the multiples of 10 and 11. Ask if pupils can find a relationship between the tables or a quick way of checking products.

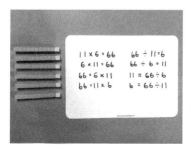

Thursday

Give each pair of pupils a set of objects and a whiteboard and pen.

Call out a multiple of 11 as a multiplication table fact (e.g. 6 × 11 = 66). Ask partner 1 in each pair to make an array that demonstrates this. Partner 2 records all the multiplication and division facts relating to the array.

Friday

Repeat Thursday's activity, with pupils swapping roles.

Week 4: Multiplication and division

Recall and use multiplication and division facts for the 12 multiplication table

Resources: counting stick, 100 squares, coloured pens, cubes / Dienes

Vocabulary: digit, repeated addition, more / less, plus, make, sum, total, altogether, one ... ten ... one hundred ... one thousand more / less, lots, multiplication, groups of, times, multiply, multiplication, multiplied by, multiple of, product, once, twice, three times ... ten times as (big, long, wide, etc.), repeated addition, array, row, column, double, halve, share, share equally, one each, two each, three each ..., group in pairs, threes ... twelves, equal groups of, divide, division, divided by / into, left, left over, remainder, ÷, ×, =, commutative law, commutativity

Throughout this week, Dienes can be useful for making the clear link that 12 equals 10 + 2; alternatively, give pupils cube towers, strips of coloured squares (see 'Counting'), small cubes or objects bundled in twelves.

Monday

Start by counting in sixes from 0–72 and back again, using either a counting stick or a 100 square to support.

Write 'If 5 × 6 = 30, 5 × 12 = 60' on the board. Ask pupils to prove this in as many ways as they can (e.g. using pictures, number sentences, arrays, groups, objects or number lines for repeated addition). Take photographs to display throughout the week.

Tuesday

Start by counting in sixes, from 0–96, circling numbers on either an individual or interactive 100 square. Then count in twelves from 0–96 and back again, using a different coloured pen on the same 100 square to highlight the relationship between the 6 and the 12 multiplication tables.

Wednesday

Start by counting in twelves from 0–144 and back again. Use Dienes to support the mental image of what 12 'looks like'.

Count together in twelves, forwards and backwards using the resource.

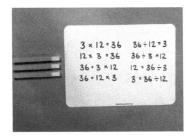

Thursday

Give each pair of pupils a set of Dienes and a whiteboard and pen.

Call out a multiple of 12 as a multiplication table fact (e.g. 3 × 12 = 36). Ask partner 1 in each pair to create a representation of this. Partner 2 records all the multiplication and division facts relating to the representation.

Friday

Repeat Thursday's activity, with pupils swapping roles.

Week 5: Multiplication and division

Solve problems involving multiplication and division, with 0 and 1

Resources: bowls, objects

Vocabulary: digit, repeated addition, more / less, plus, make, sum, total, altogether, one … ten … one hundred … one thousand more / less, lots, multiplication, groups of, times, multiply, multiplication, multiplied by, multiple of, product, once, twice, three times … ten times as (big, long, wide, etc.), repeated addition, array, row, column, double, halve, share, share equally, one each, two each, three each …, group in pairs, threes … twelves, equal groups of, divide, division, divided by / into, left, left over, remainder, ÷, ×, =, commutative law, commutativity

---

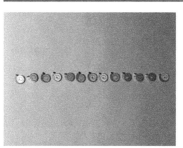

Monday

Give each pupil a bowl containing their choice of objects and a whiteboard and pen.

Remind pupils of multiplication as grouping. Write '1 × 13' on the board. Ask pupils to create an array, using any objects they like, to show this. Remind pupils that 1 group of X will always equal X. Repeat with different numbers, exploring 1 × ☐ = ☐.

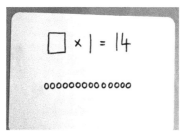

Tuesday

Give each pupil a whiteboard and pen.

Write '☐ × 1 = 14' on the board. Ask pupils to write the number that goes in the box. Remind pupils of commutativity. Ask pupils to draw an array that proves this. Repeat with other numbers.

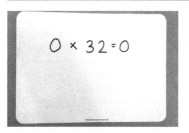

Wednesday

Give each pupil a bowl of objects and a whiteboard and pen.

Write '0 × 32 = 0' on the board. Ask pupils to create an array (they can't!), using any objects they like, to show this. Remind pupils that 0 groups of X will always equal 0.

Thursday

Remind pupils that division can be by grouping or sharing. Try sharing 12 objects among 0 people (this can be done as a whole class or in groups or pairs); how many objects does each person get? Then try placing 12 cubes into groups of 0. Try with another question. Ask pupils to work in pairs to discuss why dividing by 0 always equals 0.

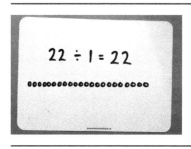

Friday

Give each pupil a bowl of objects and a whiteboard and pen.

Write '22 ÷ 1 = 22' on the board. Ask pupils to draw or create an array that matches this number sentence. Repeat with other numbers.

Recognise and use factor pairs and commutativity in mental calculations

Resources: counters

Vocabulary: digit, repeated addition, more / less, plus, make, sum, total, altogether, one ... ten ... one hundred ... one thousand more / less, lots, multiplication, groups of, times, multiply, multiplication, multiplied by, multiple of, product, once, twice, three times ... ten times as (big, long, wide, etc.), repeated addition, array, row, column, double, halve, share, share equally, one each, two each, three each ..., group in pairs, threes ... twelves, equal groups of, divide, division, divided by / into, left, left over, remainder, ÷, ×, =, commutative law, commutativity

Monday

Give each pair of pupils some counters.

Ask pupils to create an array that shows 3 × 8 = 24. Remind pupils that the array can be rotated to show that 8 × 3 = 24. Write '3 × 8 = 8 × 3' on the board.

Repeat for an array showing 7 × 4 = 4 × 7 = 28.

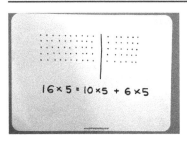

Tuesday

Give each pair of pupils a whiteboard and pen.

Ask pupils to draw an array that shows 16 × 5 = 80, using dots. Group the dots to demonstrate that 16 × 5 = 10 × 5 + 6 × 5.

Repeat for 18 × 3 (i.e. 18 × 3 = 10 × 3 + 8 × 3).

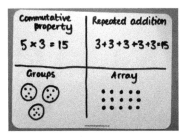

Wednesday

Give each pupil a whiteboard and pen.

Ask pupils to divide their board into four sections and label as shown. Write '5 × 3 = 15' on the board and challenge pupils to record the calculation in the four different ways. Repeat for other multiplication facts.

Thursday

Give each pupil a whiteboard and pen.

Model drawing a 'factor bug' for 12. The body has the number written on it, the legs represent the factor pairs while the head has the two factors of all integers (i.e. 1 and the number itself, 12 in this example).

Ask pupils to draw factor bugs for 24, 36, 50, 48, 60. Ask what is the same / different for each number.

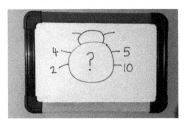

Friday

Give each pupil a whiteboard and pen.

Draw a factor bug on the board. Start to write on factor pairs and see how quickly pupils can apply their knowledge of factor pairs to calculate the number of the bug.

Week 1: Fractions

Count up and down in hundredths; recognise that hundredths arise when dividing an object by 100 and dividing tenths by 10

Resources: 10 × 10 grids of squared paper, metre sticks, colouring pencils, ten frames, objects, containers

> **Vocabulary:** whole, part, equal parts, fraction, one whole, one half, two halves, one quarter, two quarters, three quarters, four quarters, one third, two thirds, three thirds, one tenth, two tenths … ten tenths, hundredths, proportion, in every, for every, decimal, decimal fraction / point / place, numerator, denominator, equivalent, same, equal to

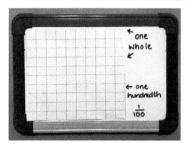

Monday

Give each pupil a 10 × 10 grid of squared paper.

Explain that each pupil has one whole block. Tell them that, when the whole block is divided into 100 equal parts, these are called hundredths. Stick a block to the board and label the parts as shown. Display throughout the week.

Together, count forwards and backwards in hundredths from 0–1 (i.e. $\frac{1}{100}$, $\frac{2}{100}$, $\frac{3}{100}$ and so on). Repeat.

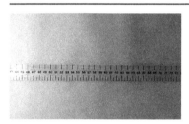

Tuesday

Give each pair of pupils a metre stick.

Explain that each pair has one whole metre. Tell them that, when the metre is divided into 100 equal parts, these are called hundredths of a metre or *centi*metres. Together, count forwards and backwards in hundredths from 0–1.

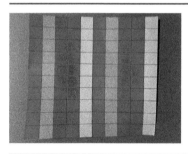

Wednesday

Give each pupil a 10 × 10 grid of squared paper.

Explain that each pupil has one whole block. Tell them that, when the block is divided into ten equal parts, these are called tenths. Ask them to colour in the tenths. Together, count forwards and backwards in tenths from 0–1. Repeat.

Explain and demonstrate that $\frac{10}{100} = \frac{1}{10}$.

Thursday

Give each pair of pupils a ten frame (or egg box, etc.) and 100 small objects (e.g. pasta, matchsticks, counters) and a whiteboard and pen.

Call out a hundredth (e.g. $\frac{85}{100}$) and ask pupils to show you that fraction using the ten frame. Encourage and model counting in tenths and hundredths. Partner 1 in each pair makes the number using the ten frame and objects. Partner 2 writes the fraction on their whiteboards.

Remind pupils of their 10 × 10 squared grids from the previous activities, making links to demonstrate that $\frac{10}{100} = \frac{1}{10}$.

Friday

Repeat Thursday's activity, with pupils swapping roles.

Week 2: Fractions

Recognise and show, using diagrams, families of common equivalent fractions

Resources: 2cm squared paper, scissors

> **Vocabulary:** whole, part, equal parts, fraction, one whole, one half, two halves, one quarter, two quarters, three quarters, four quarters, one third, two thirds, three thirds, one tenth, two tenths … ten tenths, hundredths, proportion, in every, for every, decimal, decimal fraction / point / place, numerator, denominator, equivalent, same, equal to

---

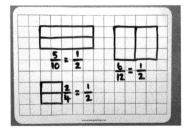

Monday

Give each pair of pupils a 1 × 6 strip of squared paper.

Ask how many squares there are in total and count them together. Explain that, when the rectangle is split into six equal parts, they are called sixths. Label. Ask pairs to draw a 1 × 6 rectangle and to split it into three equal parts, called thirds. Label. Ask pairs to draw another 6 × 1 rectangle, split it in a different way and label that fraction.

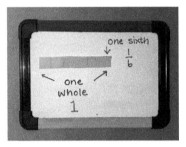

Tuesday

Give each pair of pupils a sheet of 2cm squared paper.

Ask pairs to draw one whole, halves, thirds and sixths and to compare equivalent fractions (e.g. $\frac{3}{6} = \frac{1}{2}$, $\frac{2}{3} = \frac{4}{6}$). Remind pupils that a fraction has a numerator and a denominator. The denominator tells us how many equal parts there are while the numerator tells us how many parts we have.

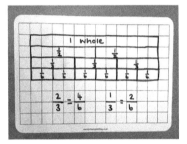

Wednesday

Give each pair of pupils a sheet of 2cm squared paper.

Ask pairs to draw one whole, thirds, quarters and twelfths. Ask pupils to find equivalent fractions (e.g. $\frac{3}{4} = \frac{9}{12}$).

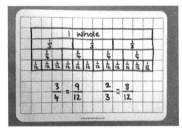

Thursday

Give each pair of pupils squared paper (or a squared whiteboard and pen).

Ask pupils to use the squares to create a 3 × 3 square as shown. Ask how many squares there are in total (count them). Ask pupils to find equivalent fractions using the square shape.

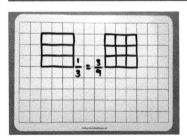

Friday

Give each pair of pupils squared paper (or a squared whiteboard and pen).

Ask pupils to find as many ways as possible to draw $\frac{1}{2}$. Repeat with other fractions (e.g. $\frac{3}{4}$).

Week 3: Fractions

Add and subtract fractions with the same denominator

Resources: cubes

Vocabulary: whole, part, equal parts, fraction, one whole, one half, two halves, one quarter, two quarters, three quarters, four quarters, one third, two thirds, three thirds, one tenth, two tenths … ten tenths, hundredths, proportion, in every, for every, decimal, decimal fraction / point / place, numerator, denominator, equivalent, same, equal to

Monday

Give each pair of pupils 3 towers of 4 cubes.

Remind pupils that a fraction has a numerator and a denominator. The denominator tells us how many equal parts there are while the numerator tells us how many parts we have.

Explain that they have 3 whole towers. Count together, in quarters, from 0 to 3 and backwards (one quarter more / less).

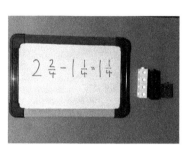

Tuesday

Give each pair of pupils 3 towers of 4 cubes.

Ask partner 1 in each pair to create the mixed number $2\frac{2}{4}$, using the cubes.

Then ask partner 2 to take away $1\frac{1}{4}$. Model writing the sentence

'$2\frac{2}{4} - 1\frac{1}{4} = 1\frac{1}{4}$'.

Repeat for other additions or subtractions using quarters.

Wednesday

Repeat Tuesday's activity, with pupils swapping roles.

Thursday

Give each pair of pupils 3 towers of 5 cubes.

Ask partner 1 in each pair to create the mixed number $2\frac{3}{5}$, using the cubes.

Then ask partner 2 to take away $1\frac{2}{5}$. Model writing the sentence

'$2\frac{3}{5} - 1\frac{2}{5} = 1\frac{1}{5}$'.

Repeat for other additions or subtractions using fifths.

Friday

Repeat Thursday's activity, with pupils swapping roles.

Week 4: Fractions

Recognise and write decimal equivalents of any number of tenths or hundredths

Resources: 100 squares

> **Vocabulary:** whole, part, equal parts, fraction, one whole, one half, two halves, one quarter, two quarters, three quarters, four quarters, one third, two thirds, three thirds, one tenth, two tenths … ten tenths, hundredths, proportion, in every, for every, decimal, decimal fraction / point / place, numerator, denominator, equivalent, same, equal to

Monday

In pairs, give each pupil a whiteboard and pen.

Ask pupils to draw a place-value grid as shown.

Call out a fraction (e.g. six tenths). Partner 1 in each pair writes it in decimal form and partner 2 writes it as a proper fraction. Repeat with pupils swapping roles.

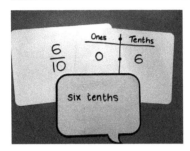

Tuesday

In pairs, give each pupil a blank 100 square, as a visual reference, and a whiteboard and pen.

Ask pupils to draw a place-value grid as shown.

Call out a fraction (e.g. sixty-five hundredths). Partner 1 in each pair writes it in decimal form and partner 2 writes it as a proper fraction.

Wednesday

Repeat Tuesday's activity, with pupils swapping roles.

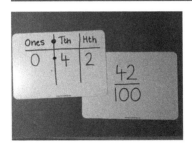

Thursday

Give each pupil a whiteboard and pen.

Write a decimal fraction (e.g. 0.42) on the board. Ask pupils to write it as a proper fraction. Repeat with other decimal fractions.

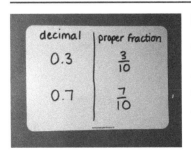

Friday

Give each pupil a whiteboard and pen.

Ask pupils to split their whiteboards in half, labelling one side 'decimal' and the other side 'proper fraction' as shown. Say fractions as tenths (e.g. three tenths) and ask pupils to write both proper fractions and decimal equivalents.

Repeat for hundredths.

Week 5: Fractions

Recognise and write decimal equivalents of any number of tenths and hundredths, and to $\frac{1}{4}$, $\frac{1}{2}$, $\frac{3}{4}$

Resources: 100 squares / squared paper, cubes

> **Vocabulary:** whole, part, equal parts, fraction, one whole, one half, two halves, one quarter, two quarters, three quarters, four quarters, one third, two thirds, three thirds, one tenth, two tenths … ten tenths, hundredths, proportion, in every, for every, decimal, decimal fraction / point / place, numerator, denominator, equivalent, same, equal to

Monday

Give each pupil a whiteboard and pen.

Call out a fraction in quarters or halves. Ask pupils to draw that fraction in any form they choose. Can they represent the fraction in four different ways?

Tuesday

Give each pupil a whiteboard and pen.

Show a blank 100 square (or prepare a 10 × 10 grid of squared paper cut into four sections of 25 squares as shown). Explain how 25 hundredth squares equals one quarter, and practise writing the decimal (0.25) and proper fraction ($\frac{1}{4}$).

Show $3\frac{3}{4}$ towers of 4 cubes (as shown). Ask pupils to write the mixed number you have shown as a decimal fraction. Repeat with other fractions.

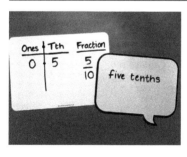

Wednesday

Give each pupil a whiteboard and pen.

Ask them to draw a place-value grid as shown.

Write 'five tenths' in words on the board. Then model writing it as a fraction $\frac{5}{10}$ and as a decimal 0.5, and show how $\frac{5}{10}$ is written on the place-value grid.

Call out a proper fraction and ask pupils to write it as a proper fraction and in decimal form. Repeat.

Thursday

Give each pair of pupils a 10 × 10 square grid (laminated, ideally, so it can be re-used, or the 10 × 10 grid as 10 strips of 10) and a whiteboard and pen each.

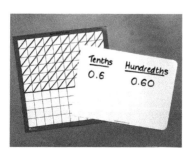

Say a number between 0 and 1 that is a multiple of a tenth (e.g. six tenths). Partner 1 in each pair makes that number using the strips of 10 squares or colouring in the 10 × 10 grid. Partner 2 writes it as a decimal fraction, in tenths and hundredths. Repeat.

Friday

Repeat Thursday's activity, with pupils swapping roles.

Week 6: Fractions

Round decimals with one decimal place to the nearest whole number

Resources: number lines (or mini counting sticks), playing cards

Vocabulary: whole, part, equal parts, fraction, one whole, one half, two halves, one quarter, two quarters, three quarters, four quarters, one third, two thirds, three thirds, one tenth, two tenths … ten tenths, hundredths, proportion, in every, for every, decimal, decimal fraction / point / place, numerator, denominator, equivalent, same, equal to

---

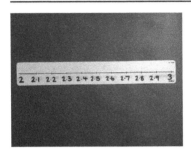

Monday

Give each pupil an empty number line with ten intervals marked (or a mini counting stick).

Count forwards and backwards from 2 to 3 in tenths.

Ask pupils to mark 2 and 3 on the number line and also tenths (in decimal form). Ask if 2 or 3 is the nearest whole number to 2.7. Repeat with other numbers. Explain that this is rounding to the nearest whole number.

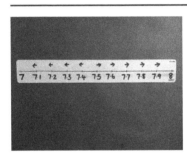

Tuesday

Give each pupil an empty number line with ten intervals marked.

Count forwards and backwards from 7 to 8 in tenths.

Ask pupils to mark 7 and 8 on the number line and also the tenths (in decimal form). Ask pupils to draw ← above all the numbers that round down to 7 and → above those that round up to 8. Tell pupils that 0.5 is a mid-point and so always rounds up. Ask if 7.4 is nearer to 7 or 8. Repeat with other numbers. Explain that this is rounding to the nearest whole number.

Wednesday

Give each pair of pupils a pack of playing cards without the picture cards and 10s, and a whiteboard and pen.

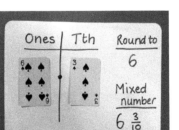

Ask pupils to write headers and mark a decimal point on their whiteboards as shown.

Partner 1 in each pair places two cards either side of the decimal point. Partner 2 rounds to the nearest whole number. Partner 1 then records the decimal as a mixed number.

Thursday

Repeat Wednesday's activity, with pupils swapping roles.

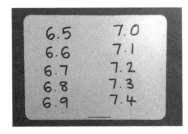

Friday

Give each pupil a whiteboard and pen.

Tell pupils you have rounded a decimal with one place to the nearest 1 and the answer was 7. What are the possible numbers? Repeat with other numbers.

Vocabulary cards

Notes